EAL2 Design
Pocket Reference

EAL2 Design
Pocket Reference

Mike Boberski

EAL2 Design Pocket Reference
by Mike Boberski

Copyright © 2006 Mike Boberski All rights reserved.
Printed in the United States of America.
Printed by Instantpublisher.com

Cover Designer: Mike Boberski
Interior Designer: Mike Boberski

Printing History:

        September 2006:   First Edition

While every precaution has been taken in the preparation of this book, the author assumes no responsibility for errors or omissions, or for damages resulting from the use of information contained herein.

ISBN: 1-59872-599-8

Table of Contents

CHAPTER 1 – INTRODUCTION

Evaluation and design evidence 1

Design evidence and evaluation strategy 3

Design evidence strategy 6

CHAPTER 2 – STRUCTURE

Describing the overall architecture 7

Physical boundaries 10

Defining subsystems 12

Describing subsystems 14

Identifying security-relevant subsystems 16

Identifying non security-relevant subsystems 18

Describing the required IT environment 19

Describing the functions provided by the IT environment 21

Describing any components involved in encryption 23

CHAPTER 3 – BEHAVIOR

Describing product behavior 27

Describing security functions 28

Describing subsystem interfaces 30

Identifying external interfaces 31

Identifying internal interfaces 32

Interface example 33

External interface description 37

CHAPTER 4 – MAPPING

Mapping interfaces to requirements 41

Mapping audit requirements example 45

Mapping audit and management requirements example 47

Mapping user data protection requirements example 48

Mapping identification and authentication requirements example 49

Mapping security function management requirements example 50

Mapping user management requirements example 51

Mapping role requirements example 52

Mapping self protection requirements example 53

APPENDIX A – ASSURANCE REQUIREMENTS

Informal functional specification 55

Descriptive high-level design 57

Informal correspondence demonstration 59

APPENDIX B – EVALUATION METHODOLOGY

Functional specification evaluation 61

High-level design evaluation...63

APPENDIX C – SECURITY TARGET REUSE

Correspondence evaluation 65

TOE description..67

TOE summary specification 69

INDEX 75

Chapter 1
Introduction

In this chapter:

Evaluation and design evidence

Design evidence and evaluation strategy

Design evidence strategy

Evaluation and design evidence

Common Criteria testing is a largely academic exercise that consists of an independent third party examining the architecture in general, and the security-related features specifically, of your IT product. Common Criteria testing also includes examining how the product is built and delivered to customers.

Common Criteria testing is called "trusted product evaluation" or simply "evaluation". It is a necessary exercise if you want to sell your product to certain Government customers. Testing is performed by an accredited commercial testing laboratory; the Government provides oversight while testing is in progress and reviews the results of testing.

Note:

 The Government, <u>not</u> the laboratory, issues a certificate when testing is complete.

Design documentation that is written for use in a Common Criteria evaluation describes your product's architecture, focusing on the security-related features. It is written by you (or a consultant on your behalf) and submitted as evaluation evidence to the testing laboratory for analysis.

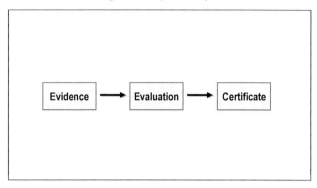

Figure 1. The Common Criteria evaluation process.

Design documentation is one document out of many that you need to write and submit to the testing laboratory to obtain Common Criteria certification. Documentation for a Common Criteria evaluation at an EAL2 level includes:

- A Security Target document
- Configuration management documentation
- Delivery and operation documentation
- Development (design) documentation
- Guidance documents documentation
- Tests and test documentation
- Vulnerability assessment documentation

"EAL2" refers to "Evaluation Assurance Level 2", which indicates how in-depth testing will be for your product.

The security-related testing requirements that your product will be tested against are defined in the Security Target. The Security Target also defines what components of your product will be tested, as well as describes how the components provide the security-related functionality to meet the requirements.

The configuration management documentation describes how you use source control tools such as Concurrent Versioning System (CVS).

The delivery and operation documentation describes how your product is delivered to the customer. The delivery and operation documentation describes how your customer knows they have received the version of the product that has been evaluated, i.e. the version that they will have expected to receive.

The development documentation is simply the design documentation, which is the subject of this reference guide.

The test documentation describes how you intend to demonstrate your product's security-related functionality. Tests can be either manually-performed or automated. Tests are used to demonstrate security-related functionality that is claimed in the Security Target using the product features described in your design documentation.

The vulnerability assessment documentation describes how you searched for obvious vulnerabilities and flaws in your product as part of its development.

Design evidence and evaluation strategy

Design evidence (documentation) should be produced as part of an overall evaluation strategy. The documents that you need to write and submit to the testing laboratory to obtain a Common Criteria certification can be sorted into two sets:

- Critical path documents
- Secondary documents

The set of documents that form a critical path from start to finish in an evaluation are those documents which must be produced in order and which progress must be made if the evaluation is ever to complete either at all or in a reasonable amount of time. The critical path documents that must be produced in order consist of the following:

1. Security Target document
2. Development (design) documentation
3. Tests and test documentation

The above documents, or at least first drafts of the above documents, must be produced in order. The Security Target defines the requirements that your product will be tested against. The design describes the structure and behavior of your product according to the requirements. The tests and test documents demonstrate how your product behaves in accordance with the requirements. The critical path through an evaluation is depicted below.

Note:

♪ If you have never gone through the process of getting a product evaluated, it is in your best interest as the product vendor to obtain professional consulting services from a reputable consultant, i.e. ask for references. <u>The information in this guide will help you work with the consultant.</u> This guide will help you understand the key concepts but it is not a comprehensive guide or a cookbook.

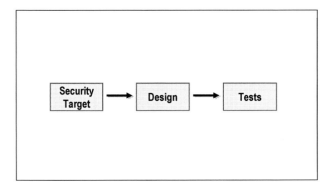

Figure 2. The critical path through an evaluation.

The rest of the documents that you need to write and submit to the testing laboratory to obtain a Common Criteria certification can simply be considered secondary documents. These documents are not optional and must be written, however they do not require the same level of effort as critical path documents, nor do they require the same level of knowledge of your product's architecture, nor do they have to be produced in a particular order. The secondary documents consist of the following:

- Configuration management documentation
- Delivery and operation documentation
- Guidance documents documentation
- Vulnerability assessment documentation

Note:

♪ In general, problems with secondary documents tend not to be show-stoppers with respect to overall evaluation progress, compared to problems with critical path documents which can stall an evaluation indefinitely.

Design evidence strategy

The purpose of this reference guide is to provide the key information that you must know to effectively write design documentation that is suitable to submit for evidence as part of a Common Criteria evaluation, regardless of overall evaluation strategy. This guide is focused on the design portion of the critical path, as depicted below.

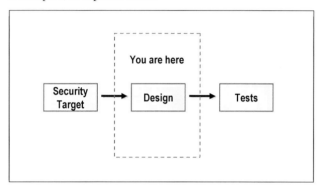

Figure 3. Design documentation is the subject of this reference guide.

Your design documentation must contain certain types of information. Regardless of your approach from a technical perspective, your design documentation must include the following types of information:

- A description of the structure of your product
- A description of the behavior of your product
- A mapping of your product's interfaces to security functions

The remainder of this guide describes each of these types of information.

Chapter 2
Structure

In this chapter:

Describing the overall architecture

Physical boundaries

Defining subsystems

Describing subsystems

Identifying security-relevant subsystems

Identifying non security-relevant subsystems

Describing the required IT environment

Describing the functions provided by the IT environment

Describing any components involved in encryption

Describing the overall architecture

You will need to describe the overall architecture of your product. Specifically, you will need to describe what is called the "evaluated configuration" of your product. Providing an overall description of the architecture of the evaluated configuration of your product is a required part of writing Common Criteria design documentation.

Before you can begin work to describe the evaluated configuration of your product, you will first need to have a "big picture" type of understanding of your product's architecture. You will also need to identify those things which are not part of your product but that your product depends on (such as an operating system if your product were to be an application).

Recycling existing documentation:

↻ The TOE architecture section of your Security Target can be reused with little or no modification to identify your product's components and components that are not part of your product but that your product depends on.

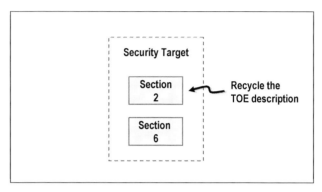

Figure 4. The TOE description (including the subsection describing physical boundaries) from your Security Target should be recycled when describing your product's structure.

Note:

♪ Avoid marketing terminology just as you would in the TOE description section of a Security Target.

The first step to describe the overall architecture is to identify your product's components. The next step is to identify components that are not part of your product but that your product depends on.

For example, the diagram below depicts an example product, let's call it "My Sample Product", that we can use in this chapter and the next to discuss subsystems. The diagram depicts servers

A and B which are product components, as well as components that are not part of the product but that the product relies on to function such as the web browser.

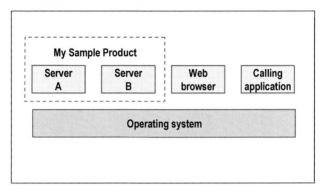

Figure 5. A sample product which consists of two server applications. Also depicted is a web browser, a calling application, and an operating system that the sample product depends on.

The diagram above depicts *My Sample Product's* components. The diagram also depicts components that are not part of the product but that the product depends on:

- **My Sample Product Server A** – This a server application that provides the *My Sample Product* functionality that end users can access.
- **My Sample Product Server B** – This is also a server application, but it is used to manage server A, for example to stop and start it and to set permissions on it.
- **My Sample Product API** – Not depicted in the figure for clarity, an API that is provided as part of *My Sample Product* to access the server
- **Web browser** – Administrative users of *My Sample Product* use a web browser to manage the server.

- **Calling application** – End users call an API that is provided as part of *My Sample Product* to access the server.
- **Web server** – Not depicted in the figure for clarity, a web server that is not part of *My Sample Product* is used by server B to provide a web-based GUI interface to administrators.
- **Operating system** – Server application A, the web server that provides a runtime environment for server B, the web browser, and the calling application each run in the context of an underlying operating system. Potentially separate operating system types and versions are not depicted in the figure, for clarity.

The above list of components can be used to describe how your product works. A narrative or description that refers to the components can be written once the components are identified, in other words. The above list of components will in any case need to be used do define both the physical and the logical boundaries of your product, as described in the next section below.

A narrative description of how your product provides its functionality that is specific to its product type is one way to provide a brief overview of your product, for example.

Quick checklist:

✓ Make sure that all of your product's parts are identified.

✓ Make sure that everything that your product needs to work is identified.

Physical boundaries

You will need to describe the physical boundaries of your product. Specifically, you will need to provide a description of the physical boundaries of the evaluated configuration of your product. Describing physical boundaries is a required part of writing Common Criteria design documentation.

Before you can begin work to describe physical boundaries, you will first need to have already at least taken a first attempt at identifying your product's components. You will also need to have already taken a first attempt at identifying components that your product needs to work.

Recycling existing documentation:

↻ The physical boundaries portion of the TOE architecture section of your Security Target can be reused with little or no modification to describe the physical boundaries of the evaluated configuration of your product.

For example, the components that make up the *My Sample Product* TOE using the example above:

- My Sample Product Server A
- My Sample Product Server B
- My Sample Product API

Continuing with the above example, the components that *My Sample Product* depends on include:

- Web browser
- Calling application
- Web server
- Operating system

Quick checklist:

✓ Make sure that all of your product's parts are identified separately from everything that your product needs to work.

Defining subsystems

You will need to describe your product's parts. Specifically, you will need to provide a description of your product's parts grouped into abstract components called "subsystems". Describing subsystems is a required part of writing Common Criteria design documentation.

Before you can begin work to group your product's parts, you will need to first have identified the physical boundaries of the evaluated configuration of your product.

Recycling existing documentation:

↻ You will <u>not</u> likely be able to recycle any existing documentation to describe your product's parts in terms of subsystems. Maybe, some existing overall architecture documentation might lend itself towards being used as a starting point.

To group your product's parts, you will first have to figure out which parts (or more precisely, which interfaces of which parts) you're going to need to demonstrate to the laboratory during testing activities near the very end of your product's evaluation.

For example, the diagram below depicts components for the example product that was used in the previous section discussing boundaries including:

- My Sample Product Server A
- My Sample Product Server B
- My Sample Product API (not depicted for clarity)
- Web browser
- Calling application
- Web server (not depicted for clarity)
- Operating system

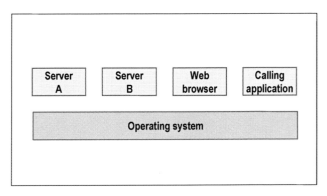

Figure 6. A sample product which consists of two server applications. Also depicted is a web browser, a calling application, and an operating system that the sample product depends on.

These components can be grouped into subsystems as follows:

- **Subsystem A** – This subsystem includes the following components:
    - My Sample Product Server A
    - My Sample Product API
- **Subsystem B** – This subsystem includes the following components:
    - My Sample Product Server B

Components grouped into subsystems are depicted in the figure below.

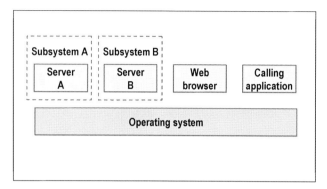

Figure 7. Two sample subsystems, a web browser, a calling application, and an operating system.

In this example, server application A and its corresponding API that can be used to access services provided by the server are grouped into a single subsystem. Similarly, server application B that can be used to manage services provided by server application A is grouped into its own separate subsystem. Components which are grouped into subsystems are depicted in the figure above.

Quick checklist:

✓ Make sure that all of your product's components are grouped into subsystems.

✓ Make sure that components grouped into a subsystem perform related functionality (at least at a fairly high level).

Describing subsystems

You will need to provide a narrative description of your product's parts after they are grouped into subsystems. It is not enough to only identify subsystems. Subsystems must be described (i.e. not just identified) as part of writing design documentation that is suitable to submit for evidence as part of a Common Criteria evaluation.

Before you can begin work to describe your product's parts, you will first need to figure out how to explain how the grouped components work together to provide functionality at high level.

Recycling existing documentation:

↻ You may be able to reuse portions of the TOE architecture section of your Security Target as a starting point to describe a subsystem or two, if there is a description of the functionality expected of a product of your type..

To describe your product's parts grouped into subsystems, you will first have to write narrative descriptions that provide an overview of the structure of the subsystem, as well as an overview of the subsystem's behavior in general terms. A subsystem description may not necessarily be longer than a couple paragraphs. One way to do this is to start subsystem description paragraphs reading something like: "When a user sends a request, the [*your subsystem's name here*] subsystem first…"

Subsystem descriptions should focus on functionality that is specific to your product's product type. For example if your product is a firewall of some sort. There should be at least one subsystem that mediates network traffic. The subsystem description for the subsystem that mediates network traffic should start off reading "When a system on an internal network initiates a connection with the firewall engine subsystem, first…". The subsystem is not described using marketing terminology. The subsystem is not described using for example UML (Unified Modeling Language). The subsystem is described in sentences, in narrative English language descriptions.

Quick checklist:

✓ Make sure that the high-level structure and organization of the subsystems is described.

✓ Make sure that the purpose of each subsystem is described.

✓ Make sure that each subsystem's interface type is described.

Identifying security-relevant subsystems

You will need to identify which of your product's subsystems are providing security-related services. It is not enough to only identify and describe subsystems, in general. Subsystems that are providing security-related services must be specifically identified as such. Identifying security-relevant subsystems is a required part of writing design documentation that is suitable to submit for evidence as part of a Common Criteria evaluation.

Before you can begin work to identify which of your product's subsystems are providing security-related services, you will first need to map your subsystem's interfaces to security functional requirements. This is described in the Chapter "Mapping", which summarizes how subsystems can be mapped to security functions.

Note:

♪ The above is not a typo. Writing design documentation that is suitable to submit for evidence as part of a Common Criteria evaluation is not a linear process, just as writing a Security Target is not. Neither document can be written start to finish. Both documents require jumping around within the document to write it, in other words.

Recycling existing documentation:

↻ It is <u>completely unlikely</u> that you will be able to recycle any existing documentation to identify which of your product's subsystems are providing security-related services.

For example, using the example product that was used in the previous section:

- **Subsystem A** – This subsystem includes the following components:
    o Supports the Security Audit security function
    o Supports the User Data Protection security function
- **Subsystem B** – This subsystem includes the following components:
    o Supports Security Management security function

Quick checklist:

✓ Make sure that subsystems that are providing security functions are mapped to security functions (not security functional requirements).

✓ Make sure that the summary of the mapping of your subsystem's interfaces to security functional requirements matches the actual mapping of your subsystem's interfaces to security functional requirements.

Identifying non security-relevant subsystems

You will need to identify which of your product's subsystems are *not* providing security-related services (if there are any). It is not enough to only identify and describe subsystems, in general. Subsystems that are not providing security-related services must be specifically identified as such. Identifying non security-relevant subsystems is a required part of writing design documentation that is suitable to submit for evidence as part of a Common Criteria evaluation.

Before you can begin work to identify which of your product's subsystems are not providing security-related services (if there are any), you will first need to identify which of your product's subsystems *are* providing security-related services, to differentiate between each set.

Recycling existing documentation:

↻ It is <u>completely unlikely</u> that you will be able to recycle any existing documentation to identify which of your product's subsystems are providing security-related services.

For example, one way to do this is to simply list the names of your product's subsystems that are not providing security-related services and identifying them as "not security relevant". If all of your product's parts are providing security-related services, you'll need to at least include along with your list of security-relevant subsystem names an additional statement something to the effect that there are no subsystems that are not providing security services.

Quick checklist:

✓ Make sure subsystems that are not providing security-related services (if there are any) are identified and labeled as such.

Describing the required IT environment

You will need to identify those things which are not part of your product but that your product depends on. For example, an operating system would be a component that an executable application will depend on. Describing the required IT environment is a required part of writing Common Criteria design documentation.

Before you can begin work to identify things that your product depends on, you will first need to figure out if there are any Security Functional requirements for the IT environment in the Security Target. This is necessary in order to figure out what requirements and what components will need to be described.

Recycling existing documentation:

↻ The list of security functional requirements for the IT environment from the security functional requirements section of your Security Target can be reused to identify the applicable requirements for the IT environment.

For example, using the example product that was used in the previous section:

If the your product as described in the Security Target were to rely on the IT environment to protect the audit trail and to protect your product in general, the list of security functional requirements might include:

- FAU_STG.1
- FPT_RVM.1
- FPT_SEP.1

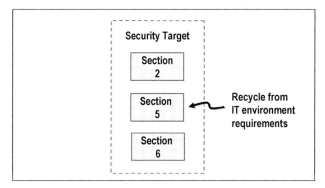

Figure 8. The security functional requirements for the IT environment from your Security Target should be recycled when describing the required IT environment.

The list of those things which are not part of your product but that your product depends on might include:

- Operating system
- Web server
- Web browser

Identifying those things which are not part of your product but that your product depends on can be done by simply listing the Security Functional Requirements for the environment as they are defined in your Security Target and by listing the things your product depends on.

Quick checklist:

✓ Make sure that if there are security functional requirements for the IT environment, that they are listed.

✓ Make sure if there are security functional requirements for the IT environment, that the components that are supporting the corresponding security functions are identified.

Describing the functions provided by the IT environment

Note:

♪ This section is <u>only</u> applicable if there are any security functional requirements for the IT environment.

You will need to describe functions provided by the IT environment. Specifically, you will need to provide a description of the security services provided by those things which are not part of your product but that your product depends on. Describing the functions provided by the IT environment is a required part of writing Common Criteria design documentation.

Recycling existing documentation:

↻ The rationale for the security functional requirements portion of your Security Target can be reused to describe the security services provided by those things which are not part of your product but that your product depends on.

Note:

♪ With the exception of cryptography as described in the next section, the one-liners from the rationale for the security functional requirements for the IT environment that are mapped to corresponding objectives will be sufficient.

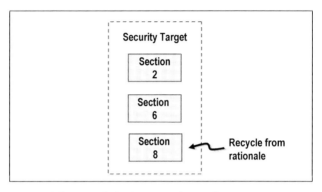

Figure 9. The rationale for the security functional requirements for the IT environment from your Security Target should be recycled when describing the functions provided by the IT environment.

For example, using the example product that was used in the previous section:

The TOE depends on the following to provide security-relevant services:

- Operating system

Operating systems in the IT environment are relied on to provide the following security-relevant services:

- FAU_SAR.1: The IT environment is required to provide interfaces to read from the audit trail generated by the TOE.

- FAU_STG.1: The IT environment is required to protect the audit trail generated by the TOE.
- FPT_RVM.1, FPT_SEP.1: The IT environment is required to protect the TOE from untrusted processes that could attempt to tamper with or bypass the TOE.
- FPT_STM.1: The IT environment is required to provide accurate and reliable time stamps

Quick checklist:

✓ Make sure if there are security functional requirements for the IT environment, that the components that are supporting the corresponding security functions are identified.

✓ Make sure that there is at least a one-liner for each security functional requirement for the IT environment.

Describing any components involved in encryption

Note:

♪ This section is only applicable if either your product or a product that yours depends on performs cryptographic functions.

You will need to describe the cryptographic services provided by *both* your product's parts as well as those things your product depends on. Describing any components involved in encryption is a required part of writing Common Criteria design documentation.

Before you can begin work to describe cryptographic services, you will first need to figure out if any cryptographic security functional requirements are claimed in the Security Target to figure out what cryptographic operation information will need to be described.

For example, if your product were to encrypt files. An example is if your product relied on a web browser and a web server that are not part of your product to support SSL.

Recycling existing documentation:

↻ The portions describing components involved in encryption in the TOE architecture section of your Security Target can be reused to describe the describe cryptographic services provided by your product and/or the environment.

Components involved in encryption need to be identified in a particular way. The following information will need to be provided for each type of cryptographic operation that is being performed by each component involved in encryption.

This information will need to be provided regardless of whether or not that component is part of your product or part of a product that yours relies on:

- Cryptographic algorithm name and applicable standard
- Cryptographic key sizes
- Cryptographic algorithm modes of operation
- Cryptographic keys and digital certificates that are used by components will also need to be identified.

For example:

Component A   This component of *My Sample Product* uses a **symmetric AES key** to encrypt files.

- This key is called the "system key"
- This key is generated by the cryptomodule in the environment
- This key is a AES 192-bit key which is generated according to FIPS-PUB 197
- This key is stored in the cryptomodule in the environment
- This key is used by the cryptomodule in environment to perform cryptographic operations
- This key is used to encrypt data using CBC mode

Quick checklist:

✓ Make sure that if you are claiming cryptographic security functional requirements in the Security Target, that you provide information about any components involved in encryption.

✓ Make sure that if you are claiming cryptographic security functional requirements, in the design you at least include placeholders that clearly identify what types of information is missing.

Chapter 3
Behavior

In this chapter:

Describing product behavior

Describing security functions

Describing subsystem interfaces

Identifying external interfaces

Identifying internal interfaces

Interface example

External interface description

Describing product behavior

You will need to describe the behavior of your product in terms of your product's parts' (i.e. subsystems') interfaces. Describing subsystem interfaces is a required part of writing Common Criteria design documentation.

Before you can begin work to describe the behavior of your product, you will first need to identify the logical boundaries of the evaluated configuration of your product. You will need to describe in a narrative how your product supports each security function, in general. Then, you will need to describe each interface to show how your product supports each security function specifically.

Recycling existing documentation:

↻ The list of security functions from the logical boundaries section of the TOE architecture section of the Security Target can be reused to figure out what security functions will need to be described and then mapped to interfaces.

Quick checklist:

✓ Make sure that there is a description for each security function that is claimed for your product in the Security Target.

Describing security functions

You will need to describe in a narrative how your product supports each security function, in general. Describing how security functions work is a required part of writing Common Criteria design documentation.

Before you can begin work to describe *how* your product supports security functions, you will first need to figure out *what* security functions will need to be described, and then provide a description for each.

Recycling existing documentation:

↻ The TOE summary specification section of your Security Target can be reused with little or no modification, for example adding "See the mapping section for more information about how this functionality is provided".

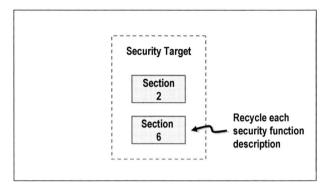

Figure 10. The security function descriptions from the TSS (TOE Summary Specification) section from your Security Target should be recycled when describing your product's security functions.

For example, one way to do this is to either recycle the TOE summary specification from your Security Target directly as described above, or at least to take the same approach that one would take in writing security function descriptions for a TOE summary specification.

Quick checklist:

✓ Make sure that there is a description for each security function that is claimed for your product in the Security Target.

✓ Make sure that the description for each security function is at least comparable to what would be found in your Security Target's TOE summary specification.

Describing subsystem interfaces

Note:

 While this section and the section below describe individual interfaces, the next Chapter ("Mapping") identifies security checks and effects that map to each of the interfaces.

You will need to describe your product's subsystems' interfaces to show how your product supports each security function. It is not enough to only describe how a subsystem provides security-related services. Individual subsystem interfaces must be described as part of writing design documentation that is suitable to submit for evidence as part of a Common Criteria evaluation.

Before you can begin work to describe your product's interfaces, you will first need to figure out what types of interfaces your subsystems provide. This will depend on the way that components were organized and grouped into subsystems to then describe individual interfaces.

Recycling existing documentation:

 You may be able to reuse certain portions of your existing API specifications or administrator guidance documentation in to describe subsystem interfaces, or at least use available interface information as a starting point.

Subsystem interfaces do not all need to be individually described. There are two types of subsystem interfaces, in general:

- External interfaces
- Internal interfaces

Each interface type is described in the following sections.

Quick checklist:

✓ Make sure that all of your subsystems' interface types (i.e. not necessarily all individual interfaces of a given type) are at least identified.

Identifying external interfaces

You will need to identify individual external interfaces, i.e. which subsystem interfaces are externally visible. It is not enough to only identify and describe subsystem interfaces. Externally-visible interfaces must also be identified as such. Identifying external interfaces is a required part of writing design documentation that is suitable to submit for evidence as part of a Common Criteria evaluation.

Before you can begin work to external interfaces, you will first need to figure out what types of interfaces your subsystems provide given the way that components were organized and grouped into subsystems to identify those which are externally visible. You will also need to figure out a reasonable approach for each interface type given the technology type.

Recycling existing documentation:

↻ It is completely unlikely that you will be able to recycle any existing documentation to identify subsystem external interface types given the way that components were organized and grouped into subsystems.

For example, end users or administrators can call external interfaces. Network traffic can be a type of external interface for a firewall or an intrusion detection type product. SQL statements and stored procedures can be external interfaces for a database type product. System calls can be external interfaces for an operating system type product. External interfaces will need to be mapped to individual security functional requirements.

Quick checklist:

✓ Make sure that all externally-visible interfaces are identified.

✓ Make sure that all externally-visible interfaces are identified in an appropriate way given the technology type.

Identifying internal interfaces

You will need to identify internal interface types. It is not enough to only identify and describe subsystem interfaces. Interfaces that are not visible outside of your product must also be identified as such. Identifying which subsystem interfaces and/or interface types are not externally-visible is a required part of writing design documentation that is suitable to submit for evidence as part of a Common Criteria evaluation.

Before you can begin work to identify internal interface types, you will first need to figure out what types of interfaces your subsystems provide. This will depend on the way that components were organized and grouped into subsystems. You will need to identify where the interfaces are located with respect to subsystems

Recycling existing documentation:

↻ It is <u>completely unlikely</u> that you will be able to recycle any existing documentation to identify subsystem internal interface types given the way that components were organized and grouped into subsystems.

For example, internal interfaces are all other subsystem interfaces besides those which are external interfaces. These interfaces are not externally visible. These interfaces cannot be called using any API, network socket, or interpreted (or compiled for that matter) language type interface. Internal interfaces will not need to be mapped to individual security functional requirements.

Quick checklist:

✓ Make sure that the location (with respect to subsystems) of each set of internal interfaces is identified.

✓ Make sure that the type of each set of internal interfaces is identified.

✓ Make sure that internal interfaces are identified as such.

Interface example

The diagram below depicts both external and internal interfaces for the example product that was used in the previous chapter discussing subsystems.

There are to recap the previous chapter (and to provide some additional detail for purposes of describing interfaces) the following components that are depicted:

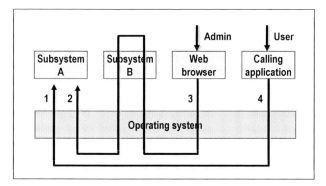

Figure 11. External and internal subsystem interfaces.

- **My Sample Product**, which is made up of the following subsystems:
    - **Subsystem A** – This subsystem is a server application that provides a network socket interface that supports a proprietary protocol. Users call an API that is provided as part of *My Sample Product* and that is part of subsystem A to access the server.
    - **Subsystem B** – This subsystem is also a server application, but it is used to manage the server in subsystem A, for example to stop and start it and to set permissions on it. This subsystem, unlike subsystem A, relies on a web server (not depicted in the figure, for clarity) that is not part of *My Sample Product* to provide a web-based GUI interface to administrators.
- **Web browser** – Administrative users of *My Sample Product* use a web browser to manage the server. Administrators access subsystem B using their web browsers; subsystem B in turn sends management requests to subsystem A using a second network

socket interface that subsystem A provides which also implements a proprietary protocol.

- **Calling application** – End users call an API that is provided as part of *My Sample Product* and that is part of subsystem A to access the server. The API initiates a network connection with subsystem A, and provides API type interfaces that wrap (i.e. abstract) individual network protocol messages.
- **Web server** – Not depicted in the figure for clarity, a web server that is not part of *My Sample Product* is used by subsystem B to provide a web-based GUI interface to administrators.
- **Operating system** – The server application of subsystem A, the web server that provides a runtime environment for subsystem B, the web browser, and the calling application each run in the context of an underlying operating system. Potentially separate operating system types and versions are not depicted in the figure, for clarity.

The diagram above depicts the following types of interfaces:

Interface #1 The network protocol interface between the calling application API and the server in subsystem A is an **internal interface**. Recall from the previous chapter that the API was included inside the TOE boundary.

Interface #2 The network protocol interface between the servers in subsystems A and B is an is an **internal interface**. Recall from the previous chapter that subsystem A is configured in the evaluated configuration only to accept network connections for this second proprietary protocol that is used for management from the machine where subsystem B is installed.

Interface #3   The web browser interface that is provided by subsystem B using the web server is an **external interface**. The interface consists of the web pages that present a GUI type interface to administrators.

Interface #4   The calling application API interface that initiates a network connection with subsystem A, and provides API type interfaces that wrap (i.e. abstract) individual network protocol messages, is an **external interface**.

The internal interfaces depicted above are not externally visible. Identifying each interface, its type, its location, and identifying it as an internal interface is typically the extent to which this interface will have to be documented. For example, in the case of interface #1 above:

- Interface #1 is called the subsystem A network interface.
- Interface #1 is the interface between the API and the server in subsystem A.
- Interface #1 consists of a proprietary network protocol.
- Interface #1 is not externally visible.

The external interfaces depicted above are externally visible, surprisingly enough. There in general the following types of external interfaces:

- API type interfaces
- GUI type interfaces
- Language type interfaces

API type interfaces may be provided by for example a Windows DLL. GUI type interfaces may be provided by a web server that are accessed using a web browser. Language type interface may be provided by for example SQL commands that are sent after first establishing a network connection to a server.

External interface description

You will need to describe external interfaces with an amount of description that is comparable for example to a UNIX man page, or for example Javadoc documentation. Providing external interface descriptions is a required part of writing design documentation that is suitable to submit for evidence as part of a Common Criteria evaluation.

Recycling existing documentation:

↻ You <u>may</u> be able to reuse certain portions of your existing API specifications or administrator guidance documentation to start work to describe subsystem interfaces.

For example, an interface description will need to include its name, a description of its purpose or usage, parameter information, and return code or exception information. External interface security checks and/or effects can be identified elsewhere in the design, as described in the next chapter.

Describing external interfaces so that the above types of information are included regardless of interface type can be difficult. Compared to an API type interface, describing a GUI type interface or a language type interface can be done in more than one way. One way to describe a GUI type interface is to describe its structure as it is organized in an existing administrator guide. For example, in the case of interface #3 above, let's say that *My Sample Product* comes with an administrator guide that has a table of contents that looks something like this:

1. Introduction
1.1 My Sample Product concepts
1.2 References
2. System requirements
2.1 Operating system requirements
2.2 Web server requirements
2.3 Web browser requirements
3. Managing users
3.1 Creating new user accounts
3.2 Assigning users to roles
4. Managing file permissions
4.1 Setting file ACLs

And, let's say that each of the sections and subsections from 2 thru 4 above in the administrator guide describe different corresponding sets of web pages to perform the task described in the administrator guide section title.

One way to describe interface #3 above is as follows:

Interface #3 The web browser interface that is provided by subsystem B using the web server is described in terms of the corresponding sections in the administrator guide as follows:

- Managing users interfaces
- Creating new user accounts interfaces
- Assigning users to roles interfaces
- Managing file permissions interfaces
- Setting file ACLs interfaces

See the *My Sample Product* administrator guide for a description of the above interfaces.

Typically, that is the extent then to which this interface will have to be documented. Again note that external interface security checks and/or effects can be identified elsewhere in the design, as described in the next chapter.

Quick checklist:

✓ Make sure it's clear how the interface can be called (i.e. make sure that the type of interface is clear).

✓ Make sure it's clear to which subsystem each interface belongs to.

✓ Make sure the interface description includes at least a minimum (remember this is basically an API specification not a programming, i.e. usage, manual) of the necessary information.

Chapter 4
Mapping

In this chapter:

Mapping interfaces to requirements

Mapping audit requirements

Mapping audit and management requirements

Mapping user data protection requirements

Mapping identification and authentication requirements

Mapping security function management requirements

Mapping user management requirements

Mapping role requirements

Mapping self protection requirements

Mapping interfaces to requirements

You will need to map subsystems to security functions, and subsystem interfaces to individual security functional requirements. Mapping subsystems to interfaces, and interfaces to requirements is a required part of writing Common Criteria design documentation.

Before you can begin work to map subsystems and subsystem interfaces, you will first need to identify external interfaces to figure out from their descriptions if they are security relevant or not. You will next have to figure out which interfaces will need to be mapped (not all subsystem interfaces need to be mapped).

Recycling existing documentation:

☋ It is <u>completely unlikely</u> that you will be able to recycle any existing documentation to map subsystems to security functions, and subsystem interfaces to individual security functional requirements.

For example, there are to recap the previous chapter, two types of subsystem interfaces, in general:

- External interfaces (these need to be mapped)
- Internal interfaces (these are **not** mapped)

End users or administrators can call external interfaces. Network traffic can be a type of external interface for a firewall or an intrusion detection type product. SQL statements and stored procedures can be external interfaces for a database type product. System calls can be external interfaces for an operating system type product. External interfaces will need to be mapped to individual security functional requirements.

Note:

♪ While all external interfaces are not necessarily mapped to security functions or security functional requirements, all security functional requirements must be mapped to at least one external interface.

Internal interfaces are all other subsystem interfaces. These interfaces are not externally visible. These interfaces cannot be called using any API, network socket, or interpreted (or compiled for that matter) language type interface.

Note:

♪ Internal interfaces are <u>not</u> mapped to security functions or security functional requirements.

An approach to mapping is to identify your product's security checks and security effects in general by first defining them, and then mapping security checks and/or effects to each of the subsystem interfaces. The goal of mapping is to demonstrate that the requirements of the Security Target have been correctly and completely implemented. This approach to mapping consists of grouping security functional requirements in such a way that they correspond to individual security checks and effects. These grouped requirements are then mapped to individual subsystem interfaces.

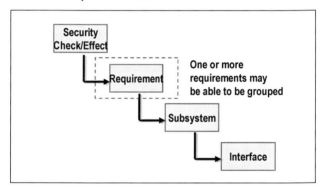

Figure 12. Mapping subsystem interfaces.

Your product performs security checks and security effects when security-relevant external interfaces are called. What makes an interface security-relevant is whether or not it performs either a security check or results in a security effect. Security checks are those things that an external interface does such as checking permissions before the caller is allowed to perform the requested function that is provided by the interface.

Security effects in general will not stop an external interface from being called, but they will result in an action such as generating an audit record in addition to performing the requested function.

Security functional requirements can be grouped in such a way that they correspond to individual security checks and effects to facilitate identifying security-relevant interfaces (and thereby facilitate mapping interfaces to security functions). For example, security functional requirements can be grouped as follows:

- FAU_GEN.1 and FAU_GEN.2
- FDP_ACC.1 and FDP_ACF.1

Interfaces that map to FAU_GEN.1 and FAU_GEN.2 generate audit records. The result of calling interfaces that map to these requirements would be a security effect, the effect being that an audit record is generated when the interface is called.

Interfaces that map to FDP_ACC.1 and FDP_ACF.1 perform an access control check. The result of calling interfaces that map to these requirements would be a security check, the check being an access control check to an object such as a file that is performed before allowing access to the file when the interface is called.

The sections that follow provide additional examples. The sections that follow also demonstrate how once security checks and security effects are defined, they can be mapped to subsystems and subsystem interfaces. Subsystems and subsystem interfaces can simply be listed for each security check or effect, since each interface would otherwise be described elsewhere in the design.

Quick checklist:

✓ Make sure that all external interfaces are mapped to security functions.

✓ Make sure that security checks and effects are defined for all mapped interfaces.

Note:

♪ Security checks and effects are <u>not</u> defined in the examples that follow.

Mapping audit requirements example

For example, mapping external interfaces to FAU_GEN.1 and FAU_GEN.2 can be done as follows:

Security Effect The following interfaces which map to FAU_GEN.1 and FAU_GEN.2 all result in the same security effect which consists of **generating an audit record**. The following interfaces all result in this effect:

- Subsystem A
  - Interface 1
  - Interface 2
  - …
- Subsystem B
  - Interface 1
  - Interface 2
  - …
- …

Mapping audit and management requirements example

For example, mapping external interfaces to FAU_SAR.1, FAU_SAR.2, FAU_SAR.3, and FMT_MTD.1 can be done as follows:

Security Effect The following interfaces which map to FAU_SAR.1, FAU_SAR.2, FAU_SAR.3, and FMT_MTD.1 all result in the same security effect which consists of **managing audit data**. The following interfaces all result in this effect:

- Subsystem A
    - Interface 1
    - Interface 2
    - …
- Subsystem B
    - Interface 1
    - Interface 2
    - …
- …

Mapping user data protection requirements example

For example, mapping external interfaces to FDP_ACC.1 and FDP_ACF.1 can be done as follows:

Security Check The following interfaces which map to FDP_ACC.1 and FDP_ACF.1 all perform the same security check which consists of **performing an access control check**. The following interfaces all perform this check:

- Subsystem A
    - Interface 1
    - Interface 2
    - …
- Subsystem B
    - Interface 1
    - Interface 2
    - …
- …

Mapping identification and authentication requirements example

For example, mapping external interfaces to FIA_UAU.1, and FIA_UID.1 can be done as follows:

Security Check   The following interfaces which map to FIA_UAU.1, and FIA_UID.1 all perform the same security check which consists of **checking that the user has been authenticated**. The following interfaces all perform this check:

- Subsystem A
    - Interface 1
    - Interface 2
    - ...
- Subsystem B
    - Interface 1
    - Interface 2
    - ...
- ...

Mapping security function management requirements example

For example, mapping external interfaces to FMT_MSA.1, FMT_MSA.3, and FMT_SMF.1 can be done as follows:

Security Effect The following interfaces which map to FMT_MSA.1, FMT_MSA.3, and FMT_SMF.1 all result in the same security effect which consists of **managing access to an object**. The following interfaces all result in this effect:

- Subsystem A
    - Interface 1
    - Interface 2
    - …
- Subsystem B
    - Interface 1
    - Interface 2
    - …
- …

Mapping user management requirements example

For example, mapping external interfaces to FMT_MTD.1 and FMT_SMF.1 can be done as follows:

Security Effect The following interfaces which map to FMT_MTD.1 and FMT_SMF.1 all result in the same security effect which consists of **changing passwords**. The following interfaces all result in this effect:

- Subsystem A
    - Interface 1
    - Interface 2
    - …
- Subsystem B
    - Interface 1
    - Interface 2
    - …
- …

Mapping role requirements example

For example, mapping external interfaces to FMT_SMR.1 can be done as follows:

Security Check The following interfaces which map to FMT_SMR.1 all perform the same security check which consists of **checking that the user is an administrator**. The following interfaces all perform this check:

- Subsystem A
  - Interface 1
  - Interface 2
  - …
- Subsystem B
  - Interface 1
  - Interface 2
  - …
- …

Mapping self protection requirements example

For example, mapping external interfaces to FPT_RVM.1 and FPT_SEP.1 can be done as follows:

Security Checks and Effects	The interfaces that map to FPT_RVM.1 and FPT_SEP.1 consist of the set of interfaces that are mapped to other security functions.
	(no individual interfaces would be listed)

Appendix A
Assurance Requirements

Note:

♪ The guidance provided in this reference is intended to help you write design documentation that meets the assurance requirements that are provided below for reference only.

Informal functional specification

The following are the applicable EAL2 security assurance requirements for the functional specification portions of the design documentation.

ADV_FSP.1.1d   The developer shall provide a functional specification.

ADV_FSP.1.1c   The functional specification shall describe the TSF and its external interfaces using an informal style.

ADV_FSP.1.2c   The functional specification shall be internally consistent.

ADV_FSP.1.3c   The functional specification shall describe the purpose and method of use of all external TSF interfaces, providing details of effects, exceptions and error messages, as appropriate.

ADV_FSP.1.4c The functional specification shall completely represent the TSF.

ADV_FSP.1.1e The evaluator shall confirm that the information provided meets all requirements for content and presentation of evidence.

ADV_FSP.1.2e The evaluator shall determine that the functional specification is an accurate and complete instantiation of the TOE security functional requirements.

Descriptive high-level design

The following are the applicable EAL2 security assurance requirements for the high-level design portions of the design documentation.

ADV_HLD.1.1d The developer shall provide the high-level design of the TSF.

ADV_HLD.1.1c The presentation of the high-level design shall be informal.

ADV_HLD.1.2c The high-level design shall be internally consistent.

ADV_HLD.1.3c The high-level design shall describe the structure of the TSF in terms of subsystems.

ADV_HLD.1.4c The high-level design shall describe the security functionality provided by each subsystem of the TSF.

ADV_HLD.1.5c The high-level design shall identify any underlying hardware, firmware, and/or software required by the TSF with a presentation of the functions provided by the supporting protection mechanisms implemented in that hardware, firmware, or software.

ADV_HLD.1.6c The high-level design shall identify all interfaces to the subsystems of the TSF.

ADV_HLD.1.7c The high-level design shall identify which of the interfaces to the subsystems of the TSF are externally visible.

ADV_HLD.1.1e The evaluator shall confirm that the information provided meets all requirements for content and presentation of evidence.

ADV_HLD.1.2e The evaluator shall determine that the high-level design is an accurate and complete instantiation of the TOE security functional requirements.

Informal correspondence demonstration

The following are the applicable EAL2 security assurance requirements for the correspondence portions of the design documentation.

ADV_RCR.1.1d The developer shall provide an analysis of correspondence between all adjacent pairs of TSF representations that are provided.

ADV_RCR.1.1c For each adjacent pair of provided TSF representations, the analysis shall demonstrate that all relevant security functionality of the more abstract TSF representation is correctly and completely refined in the less abstract TSF representation.

ADV_RCR.1.1e The evaluator shall confirm that the information provided meets all requirements for content and presentation of evidence.

Appendix B
Evaluation Methodology

Note:

♪ The guidance provided in this reference is intended to help you write design documentation that passes the scrutiny of the testing laboratory as described below for reference only.

Functional specification evaluation

The following are the applicable EAL2 evaluation requirements for the functional specification portions of the design documentation. This is the criteria that this aspect of the design will be reviewed against to determine if applicable security functional requirements have been met.

2:ADV_FSP.1-1 The evaluator shall examine the functional specification to determine that it contains all necessary informal explanatory text.

2:ADV_FSP.1-2 The evaluator shall examine the functional specification to determine that it is internally consistent.

2:ADV_FSP.1-3 The evaluator shall examine the functional specification to determine that it identifies all of the external TOE security function interfaces.

2:ADV_FSP.1-4 The evaluator shall examine the functional specification to determine that it describes all of the external TOE security function interfaces.

2:ADV_FSP.1-5 The evaluator shall examine the presentation of the TSFI to determine that it adequately and correctly describes the behaviour of the TOE at each external interface describing effects, exceptions and error messages.

2:ADV_FSP.1-6 The evaluator shall examine the functional specification to determine that the TSF is fully represented.

2:ADV_FSP.1-8 The evaluator shall examine the functional specification to determine that it is an accurate instantiation of the TOE security functional requirements.

High-level design evaluation

The following are the applicable EAL2 evaluation requirements for the high-level design portions of the design documentation. This is the criteria that this aspect of the design will be reviewed against to determine if applicable security functional requirements have been met.

2:ADV_HLD.1-1 The evaluator shall examine the high-level design to determine that it contains all necessary informal explanatory text.

2:ADV_HLD.1-2 The evaluator shall examine the presentation of the high-level design to determine that it is internally consistent.

2:ADV_HLD.1-3 The evaluator shall examine the high-level design to determine that the TSF is described in terms of subsystems.

2:ADV_HLD.1-4 The evaluator shall examine the high-level design to determine that it describes the security functionality of each subsystem.

2:ADV_HLD.1-5 The evaluator shall check the high-level design to determine that it identifies all hardware, firmware, and software required by the TSF.

2:ADV_HLD.1-6 The evaluator shall examine the high-level design to determine that it includes a presentation of the functions provided by the supporting protection mechanisms implemented in the underlying hardware, firmware, or software.

2:ADV_HLD.1-7 The evaluator shall check that the high-level design identifies the interfaces to the TSF subsystems.

2:ADV_HLD.1-8 The evaluator shall check that the high-level design identifies which of the interfaces to the subsystems of the TSF are externally visible.

2:ADV_HLD.1-9 The evaluator shall examine the high-level design to determine that it is an accurate instantiation of the TOE security functional requirements.

2:ADV_HLD.1-10 The evaluator shall examine the high-level design to determine that it is a complete instantiation of the TOE security functional requirements.

Correspondence evaluation

The following are the applicable EAL2 evaluation requirements for the correspondence portions of the design documentation. This is the criteria that this aspect of the design will be reviewed against to determine if applicable security functional requirements have been met.

2:ADV_RCR.1-1 The evaluator shall examine the correspondence analysis between the TOE summary specification and the functional specification to determine that the functional specification is a correct and complete representation of the TOE security functions.

2:ADV_RCR.1-2 The evaluator shall examine the correspondence analysis between the functional specification and the high-level design to determine that the high-level design is a correct and complete representation of the functional specification.

Appendix C
Security Target Reuse

Note:

♪ The guidance provided in this reference is intended to help you navigate sections of your Security Target that may be recycled in your design as described in this guide. The assurance requirements that are provided below are for reference only.

TOE description

The following are a (relevant) subset the applicable assurance requirements and corresponding Security Target evaluation methodology that may be helpful trying to write similar and/or corresponding sections in your design.

ASE_DES.1.1C   The TOE description shall describe the product or system type, and the scope and boundaries of the TOE in general terms both in a physical and a logical way.

ASE_DES.1-1   The evaluator shall examine the TOE description to determine that it describes the product or system type of the TOE.

The evaluator determines that the TOE description is sufficient to give the reader a general understanding of the intended usage of the product or system, thus providing a context for the evaluation. Some examples of product or system types are: firewall, smartcard, crypto-modem, web server, intranet.

There are situations where it is clear that some functionality is expected of the TOE because of its product or system type. If this functionality is absent, the evaluator determines whether the TOE description adequately discusses this absence. An example of this is a firewall-type TOE, whose TOE description states that it cannot be connected to networks.

ASE_DES.1-2 The evaluator shall examine the TOE description to determine that it describes the physical scope and boundaries of the TOE in general terms.

The evaluator determines that the TOE description discusses the hardware, firmware and software components and/or modules that constitute the TOE at a level of detail that is sufficient to give the reader a general understanding of those components and/or modules.

If the TOE is not identical to a product, the evaluator determines that the TOE description adequately describes the physical relationship between the TOE and the product.

ASE_DES.1-3 The evaluator shall examine the TOE description to determine that it describes the logical scope and boundaries of the TOE in general terms.

The evaluator determines that the TOE description discusses the IT, and in particular the security features offered by the TOE at a level of detail that is sufficient to give the reader a general understanding of those features.

If the TOE is not identical to a product, the evaluator determines that the TOE description adequately describes the logical relationship between the TOE and the product.

TOE summary specification

ASE_TSS.1.1C The TOE summary specification shall describe the IT security functions and the assurance measures of the TOE.

ASE_TSS.1-1 The evaluator shall check that the TOE summary specification describes the IT security functions and assurance measures of the TOE.

The evaluator determines that the TOE summary specification provides a high-level definition of the security functions claimed to meet the TOE security functional requirements, and of the assurance measures claimed to meet the TOE security assurance requirements.

The assurance measures can be explicitly stated, or defined by reference to the documents that satisfy the security assurance requirements (e.g. relevant quality plans, life cycle plans, management plans).

ASE_TSS.1.2C The TOE summary specification shall trace the IT security functions to the TOE security functional requirements such that it can be seen which IT security functions satisfy which TOE security functional requirements and that every IT security function contributes to the satisfaction of at least one TOE security functional requirement.

ASE_TSS.1-2 The evaluator shall check the TOE summary specification to determine that each IT security function is traced to at least one TOE security functional requirement.

Failure to trace implies that either the TOE summary specification is incomplete, the TOE security functional requirements are incomplete, or the IT security function has no useful purpose.

ASE_TSS.1.3C The IT security functions shall be defined in an informal style to a level of detail necessary for understanding their intent.

ASE_TSS.1-3 The evaluator shall examine each IT security function to determine that it is described in an informal style to a level of detail necessary for understanding its intent.

In some cases, an IT security function may provide no more detail than is provided in the corresponding TOE security functional requirement or requirements. In others, the ST author may have included TOE-specific details, for example using TOE-specific terminology in place of generic terms such as "security attribute".

Note that a semi-formal or formal style of describing IT security functions is not allowed here, unless accompanied by an informal style description of the same functions. The goal here is to understand the intent of the function, rather than determining properties such as completeness or correctness of the functions.

ASE_TSS.1.4C All references to security mechanisms included in the ST shall be traced to the relevant security functions so that it can be seen which security mechanisms are used in the implementation of each function.

ASE_TSS.1-4 The evaluator shall examine the TOE summary specification to determine that all references to security mechanisms in the ST are traced back to IT security functions.

References to security mechanisms are optional in an ST but may (for example) be appropriate where there is a requirement to implement particular protocols or algorithms (e.g. specified password generation or encryption algorithms). If the ST contains no references to security mechanisms, this work unit is not applicable and is therefore considered to be satisfied.

The evaluator determines that each security mechanism that the ST refers to is traced back to at least one IT security function. Security Target evaluation

Failure to trace implies that either the TOE summary specification is incomplete or the security mechanism has no useful purpose.

ASE_TSS.1.5C The TOE summary specification rationale shall demonstrate that the IT security functions are suitable to meet the TOE security functional requirements.

ASE_TSS.1-5 The evaluator shall examine the TOE summary specification rationale to determine that for each TOE security functional requirement it contains an appropriate justification that the IT security functions are suitable to meet that TOE security functional requirement.

If no IT security functions trace back to the TOE security functional requirement, this work unit fails.

The evaluator determines that the justification for a TOE security functional requirement demonstrates that if all IT security functions that trace back to that requirement are implemented, the TOE security functional requirement is met.

The evaluator also determines that each IT security function that traces back to a TOE security functional requirement, when implemented, actually contributes to meeting that requirement.

Note that the tracings from IT security functions to TOE security functional requirements provided in the TOE summary specification may be a part of a justification, but do not constitute a justification by themselves.

Index

A

administrator guidance 30, 37
API 9, 10, 11, 12, 13, 14, 30, 33, 34, 35, 36, 37, 39, 42
API specification 30, 37, 39
API specifications 30, 37
API type interface ... 35, 36, 37
API type interfaces .. 35, 36, 37
Assurance Requirement 55
Assurance Requirements 55

B

boundaries 10, 11, 12, 27, 28, 67, 68

C

commercial testing laboratory 1
Common Criteria 1, 2, 3, 5, 6, 7, 11, 12, 15, 16, 18, 19, 21, 23, 27, 28, 30, 31, 32, 37, 41

Common Criteria testing..................... 1
Configuration management 2, 5
Correspondence 65
Correspondence evaluation.............. 65
Critical path 3
cryptographic 23, 24, 25
cryptographic function 23
cryptographic functions 23
cryptographic services 23, 24
cryptography 22

D

Delivery and operation 2, 5
Descriptive high-level design 57
Design 1, 2, 3, 6
Development 2, 4
Development (design) 2, 4

E

EAL2 . 2, 55, 57, 59, 61, 63, 65

evaluated configuration 7, 11, 12, 27, 35
Evaluation Assurance Level 2 2
Evaluation Methodology 61
existing documentation .8, 11, 12, 15, 17, 18, 19, 21, 24, 28, 30, 31, 32, 37, 42
external and internal interface 33
external and internal interfaces 33
External interface 30, 32, 37, 42
External interfaces... 30, 32, 42
externally visible 31, 33, 36, 42, 57, 64
externally-visible...... 32
externally-visible interface 32
externally-visible interfaces 32

F

Functional requirement 19
Functional requirements 19
Functional specification 61
Functional specification evaluation 61

G

GUI ... 10, 34, 35, 36, 37
GUI type interface ...36, 37
GUI type interfaces.. 36, 37

H

High-level design 63
High-level design evaluation 63

I

Informal correspondence demonstration........ 59
Informal functional specification 55
interface type 16, 31, 32, 37
interface types 31, 32
interfaces.. 6, 12, 16, 17, 22, 27, 28, 30, 31, 32, 33, 35, 36, 37, 38, 41, 42, 43, 44, 45, 46, 47, 48, 49, 50, 51, 52, 53, 55, 57, 61, 62, 64
interfaces to requirements......... 41
internal interface 32, 33, 35, 36
Internal interface 30, 33, 42, 43

internal interfaces 32, 33, 36
Internal interfaces 30, 33, 42, 43
IT environment ... 19, 20, 21, 22, 23

L

Language type interface 36, 37
Language type interfaces 36

M

map subsystems and subsystem interfaces 41
mapping 6, 17, 28, 43, 44, 46, 47, 48, 49, 50, 51, 52, 53
Mapping audit 46, 47
Mapping audit and management 47
Mapping identification and authentication 49
Mapping role 52
Mapping security function management 50
Mapping self protection 53
Mapping subsystem. 41, 43
Mapping subsystems 41
Mapping subsystems to interfaces 41
Mapping user data protection 48
Mapping user management 51

N

narrative . 10, 14, 15, 27, 28
narrative description 10, 14, 15

P

physical boundaries ... 8, 10, 11, 12

Q

Quick checklist .. 10, 11, 14, 16, 17, 19, 21, 23, 25, 28, 29, 31, 32, 33, 39, 45

R

rationale 21, 22, 73

S

Secondary 3
security check 30, 37, 39, 43, 44, 45, 48, 49, 52
security checks ... 30, 37, 39, 43, 44, 45
security checks and effects. 30, 43, 44, 45
security effect 43, 44, 46, 47, 50, 51

security effects 43, 44
security function...6, 16, 17, 19, 20, 21, 22, 23, 24, 25, 27, 28, 29, 30, 32, 33, 41, 42, 43, 44, 45, 53, 56, 57, 58, 59, 61, 62, 63, 64, 65, 69, 70, 71, 72, 73
Security functional requirement.......... 44
Security functional requirements 44
security functions .6, 16, 17, 21, 23, 28, 29, 41, 42, 43, 44, 45, 53, 65, 69, 70, 71, 73
Security Target ...2, 3, 4, 8, 11, 15, 16, 19, 20, 21, 22, 24, 25, 28, 29, 43, 67, 72
security-related...1, 2, 3, 16, 17, 18, 19, 30
security-relevant.16, 18, 22, 43, 44
security-relevant subsystems.......16, 18
subsystem 13, 14, 15, 16, 17, 18, 27, 30, 31, 32, 34, 35, 36, 37, 38, 39, 41, 42, 43, 44, 57, 63
subsystem interface..27, 30, 31, 32, 34, 37, 41, 42, 43, 44
subsystem interfaces 27, 30, 31, 32, 34, 37, 41, 42, 43, 44
subsystems8, 12, 13, 14, 15, 16, 17, 18, 19, 27, 30, 31, 32, 33, 34, 35, 41, 42, 44, 57, 63, 64

T

technology type... 31, 32
Tests..................... 2, 3, 4
TOE ... 8, 11, 15, 22, 23, 24, 28, 29, 35, 56, 58, 61, 62, 64, 65, 67, 68, 69, 70, 71, 72, 73
TOE architecture.. 8, 11, 15, 24, 28
TOE description... 8, 67, 68, 69
TOE summary specification .. 28, 29, 65, 69, 70, 71, 72, 73
trusted product evaluation............... 1
TSS ... 29, 69, 70, 71, 73